Maquator 90
- Card Games

Play with numbers for big and small 😊

Provided by Inprudea Ltd.

2023 version

Dear Maquator Player,

In these instructions you will find an overview of the Maquator 90 games:

Game	Game Complexity	For players who know or are learning:
➢ UpDown ➢ UpDown Solitaire	Low	➢ *The numbers 0 to 9.*
➢ Duel ➢ Line 5 ➢ 45er ➢ Equate Solitaire	Medium	➢ *Basic math including the operators plus, minus, divide and multiply.*
	High	➢ *More advanced operators such as power of/index, fractions, rounding and decimals.*

How to use this guide

First choose which card game you wish to play. Then go to the instruction section for each game. Finally start playing either on your own or with your friends. Play and have fun!!!!

The Components of the Game

- This instruction booklet.
- Not included, but necessary to play:
    - A set of Maquator 90 playing cards consisting of 50 cards with numbers from 0 to 9 and with different colour coding. The ideal size of cards is small e.g. 4 cm x 6 cm. This is to allow for a better playing experience.
    - The Maquator 90 cards can be purchased separately. However, you can also create your own set of cards. See the Appendix of this booklet for the steps to design your own set of cards.

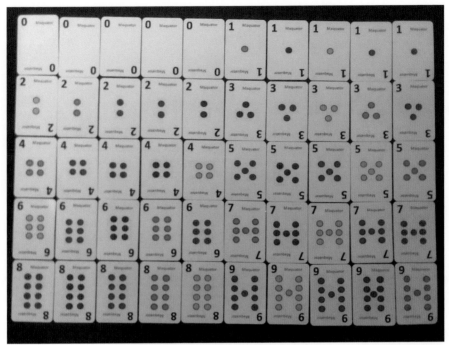

The Maquator 90 set of cards.

What do the playing cards mean

When playing *Maquator Games 90 Games* the cards 0 to 9 simply means the number written on the cards.

The cards 1-9 have been marked with dots representing the written numbers on each card. The dots are coloured to play 45er, but with colours.

For the best playing experience, it is recommended to remove 3 of the 5 cards marked 0 except when playing 45er where it is recommended to play with all 5 cards marked 0.

Card Game: UpDown

Number of players

➤ 2-5 players

Preparing to play

➤ Lay out all the playing cards on a table face down.
➤ Turn one card in the middle face up and leave the remaining cards face down. The first turned card is the ***original face up card***.

The original face up card is placed in the centre with the remaining cards face down.

What to do

➤ The aim of UpDown is to lay down cards in sequence either one more or one less of cards already lying face up and connected to the original face up card.

Maquator 90

How to play

- ➤ Decide which player starts. Each player now takes a turn. In each turn a player can turn one card face up.
- ➤ The first player turns a card lying face down. If the card is one more or one less than the original face up card then this card can be placed next to (connected to) the original face up card.
- ➤ If a card is not one more or one less than the original face up card or subsequent connected cards, then the card is left face up but not connected to other cards.
- ➤ The next player now turns a card and, if possible, connects the card to the original face up cards or subsequent connected cards…and so on.

Cards which are one more or one less than the original face up card, or subsequent connected cards, are connected.

- ➤ If a player lays down an adjacent card to the original face up card or already connected cards, the player can now <u>in the same turn</u> connect other cards showing face up, but which were not connected when the player's turn started.

> ➢ If a newly turned card cannot be added to the original face up card or already connected cards then the turn passes to the next player, who continues to turn a card.
> ➢ The game finishes when all cards have been turned and connected to the connected cards lying face up.

How to win or lose

> ➢ At the start of each game each player has *one point*.
> ➢ A point can be gained if a player connects 4 or more cards in a turn.
> ➢ A point can be lost if a player in the turn fails to connect unconnected cards facing up and which are one more or less than any already connected cards.

Note, if the player does not add the card 6 in the turn, then the player will lose a point.

> ➢ If you do not have any points left and you lose another point, then you have lost and do not continue to play.
> ➢ The player with most points at the end of the game wins or in case, where there is only one player left, the remaining player wins. If

one or more players have the same number of points at the end of the game, then it is a draw.

Variations of UpDown

> In each turn a player can turn two cards rather than one card to fasten the game.
> UpDown can also be played where all cards except the original face up card is kept in a stack of cards instead of being spread out face down. In this case the original face up card is placed on the table. Each player now turns cards from the stack of cards. If a card can not be connected to the original face up card or connected cards it is laid face up on the table (but not connected) and can be connected in future turns.
> UpDown can be played, where if a card cannot be connected to the original face card, it is placed face down on the table. When playing this way, then the players have to remember the cards, which have already been turned, but put back down, to have the best playing experience.
> You can also play UpDown on your own for fun or in teams!

The game finishes when all cards have been turned and connected.

Card Game: UpDown Solitaire

Number of players

➢ 1 player

Preparing to play

➢ Turn one card face up. The first turned card is the **original face up card**.
➢ Lay in a row 10 cards face up. These cards are called the **pool cards**.
➢ Leave the remaining cards face down in a stack within reach.

The original face up card is placed above the row of pool cards.

What to do

➢ The aim of UpDown Solitaire is to connect cards in sequence either one more or one less than cards already lying face up and connected to the original face up card or other cards.

> Cards being connected to the original face up cards or subsequent connected cards are taken from the pool cards.

How to play

> From the pool cards, place any of the cards, which are one more or one less adjacent to the original face up or subsequent cards connected to the original face up card.
> At the start of each turn ensure that there are 10 pool cards lying face up.
> Continue to build out a net of connected cards where each adjacent card is one less or one more that the one next to it.

Cards from the pool cards are connected to the original face up card and subsequent connected cards. Each round starts with 10 cards.

> If you cannot lay down a pool card, you can change one or more pool cards by placing the pool card in the bottom of the stack and replacing it with a card from the top of the stack.
> The solitaire completes once all cards in the stack and the pool cards have been connected to each other.

Pool cards connected to other cards.

A new round starts.

Maquator 90

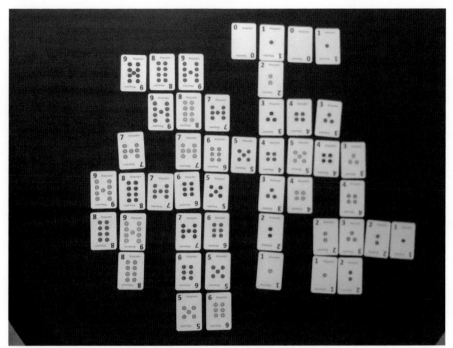

The game finishes when all cards have been connected.

Variations

> ➤ UpDown Solitaire can be played with fewer pool cards e.g. 8. It is likely that a player in that case will have to change cards from the stack of cards more frequently.

Introduction to Duel, Line 5, 45er and Equate Solitaire

The games in this section are based on the idea that a number row of playing cards on its own can represent an equation.

First, what is an equation? In this context, it is a calculation on one side of the equal sign, i.e. =, which is equal to a calculation or result on the other side of the equal sign for example 3 = 2+1 or 6-3 = 2+1.

If you take away the operator and equal signs in an equation then a row of numbers such as 321 can represent 3 = 2+1 or 123 can represent 1+2 = 3. As shown below 3 = 2 + 1.

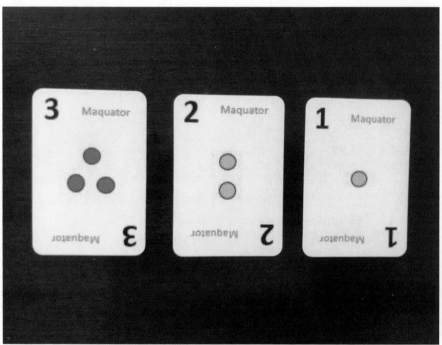

3 = 2 + 1 or 3 − 2 = 1 represented by 3 2 1.

Card Game: Duel

Number of players

> ➤ 2-5 players

Preparing to play

> ➤ Turn one card and lay this card face up in the middle of the players. The first turned card is known as the **original face up card**.
> ➤ Hand out 5 cards to each player. **A player's cards should be hidden to other players.**
> ➤ The remaining cards are placed faced down in a stack so each player can reach the stack of cards.

The row of 5 cards representing a player's hand is normally hidden but here for illustration shown face up (top and bottom row).

What to do

- ➤ The aim of the game is to lay down number rows, connected to the original face up card or other connected cards, representing mathematical equations.
- ➤ When playing Duel each new equation represented by a number row has to connect to minimum one card already placed face up.

Which operators to use

The following operators and symbols can be used when forming equations:

- ➤ + i.e. addition.
- ➤ - i.e. minus.
- ➤ * i.e. multiply.
- ➤ / i.e. divide.
- ➤ = i.e. equal sign.
- ➤ ( ) i.e. brackets - optional

How to play

- ➤ Each player takes turn to lay down a number row connecting a number row to minimum one card already laying face up.
- ➤ The first player connects the number row to the original face up card.
- ➤ The second player lays down a number row representing an equation using a card already placed by a previous player.
- ➤ The game continues by players connecting number rows using minimum one card already placed face up.
- ➤ At the start of each turn a player picks up additional cards from the stack of cards to have 5 cards facing up (or in the hand).
- ➤ All adjacent cards need to form part of a number row representing a valid equation.
- ➤ If a player cannot or chooses not to lay down a number row in a turn the player can pass and take an additional card instead.

Maquator 90

The first player adds 1 + 2 + 7 equal to 10 i.e. 1 2 7 1 0.

The second player adds 1+4+7 = 8 + 4 i.e. 1 4 7 8 4.

A new round starts each player having 5 cards.

> ➢ A player can lay down more than one number row in each turn.
> ➢ The game finishes when there are no more cards in the stack of cards and all players have had a final chance to lay down more number rows.

How to win and lose

The player with most points at the end of the game is the winner.
Points can be given using either the **Simple** point system or the **Simple+** (read Simple Plus) point system. **When playing you have to agree from the start which point system to use.**

Point System: **Simple**
> ➢ 1 point is given per card placed on the table in each turn. E.g. if four cards has been laid down then 4 points will be rewarded.

Point System: **Simple+**
> ➢ 1 point is given per card placed on the table.
> ➢ A bonus point is given depending on how many tens there are in the result of the equation i.e. **15**=… gives 1 bonus point, **47**=…

gives 4 bonus points. A maximum of 10 bonus points can be rewarded in a turn. If the equation does not have a result on either side of the equation e.g. 4 * 4 = 8 + 8, then bonus points are not rewarded.

*Example of a bridge where 4 * 15 = 60 i.e. 4 1 5 6 0. According to the Simple point system 3 points are awarded; with Simple+ then 3 + 6 = 9 points are rewarded for 3 cards laid down and 6 tens in 60.*

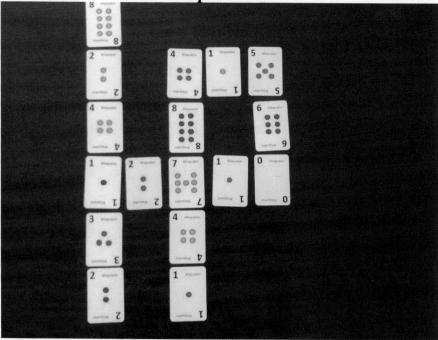

((2 * 3) + 1) * 4 = 28 i.e. 2 3 1 4 2 8. According to the Simple point system 5 points is awarded; under Simple+ then 5 + 2 points is rewarded.

Variations
Duel can be played with the following variations.

> **Advanced operators** where the players can apply:
>   o Brackets i.e. ( ).
>   o Fractions and decimals e.g. 5 / 4 = 1.25 written as 5 4 1 2 5
>   o Power of / Indices e.g. $2^2 = 4$ written as 2 2 4
>   o Mean average e.g. Average (4,5,6) = 5 written 4 5 6 5
>   o Rounding which implies rounding the result to <u>two decimal places.</u> As an example, 1+1/3 = 1.33 written as 1 1 3 1 3 3 with 1.33 rounded to two decimals.
> **Snatching.** When younger players are playing with grown-ups, the grown-ups can play with cards facing up and placed on the table in front of the grown-up player. Younger players can in their turn snatch a card from an older player thereby having a larger choice of cards to create equations from.
> When playing with Snatching, all players still start a turn with at least 5 cards.
> **Additional cards.** Duel can be played with two sets of cards allowing for a longer playing experience.

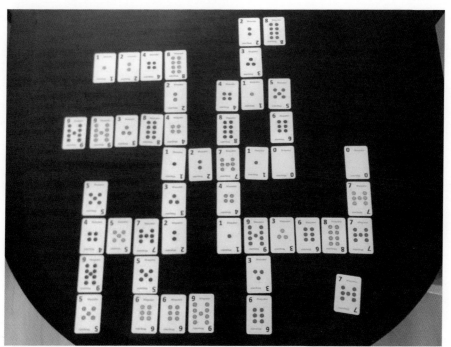

The game completes when both players cannot lay down more number rows.

Card Game: Line 5

Number of players: 2-5 players

Preparing to Play

> ➤ Lay out 5 cards on a row with face up on a table.
> ➤ Hand out 5 cards to each player. The cards on the player's hand should be hidden.
> ➤ Place the remaining stack of cards so that all players can reach the stack.

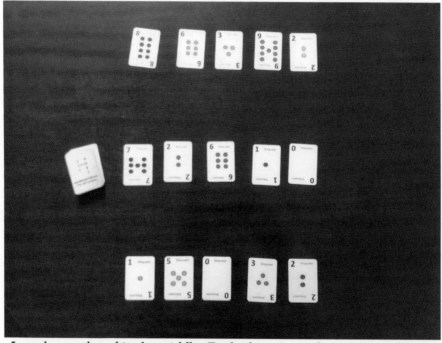

5 cards are placed in the middle. Each player's cards are shown face up top and bottom for illustration, but normally held by the players.

What to do

> ➤ The aim of the game is to create number rows representing mathematical equations using the cards placed face up on the table and/or held in the hand by the player.

Maquator 90

Which operators to use

The following operators and symbols can be used when forming equations:

> + i.e. addition.
> - i.e. minus.
> * i.e. multiply.
> / i.e. divide.
> = i.e. equal sign.
> ( ) i.e. brackets - optional

How to Play

> The first player lays down a number row from the cards held in the hand, or from the cards laying face up on the table or a combination of the two.

*In the above the player lay down 20 = 10 * 2 i.e. 2 0 1 0 2.*

> Once a number row representing an equation is created, then the number row of cards is removed and kept by the player.

21

Pool cards are added so each player starts with 5 pool cards.

- ➢ Before the next player's turn, cards are added from the stack of cards so that each turn starts with 5 cards lying face up on the table and each player is holding 5 cards.
- ➢ The next player then lays down a number row representing an equation and keeps the cards. Multiple number rows can be created in each turn.
- ➢ If a player cannot lay down a number row, the player takes another card from the stack of cards and passes the turn.
- ➢ Once there are no further cards in the stack of cards and the players cannot lay down more number rows then the game finishes.

How to win and lose

- ➢ The player with the most cards at the end of the game wins. If one or more players have the same amount of cards, it is a draw.

Maquator 90

The player lays down (6 / 2) + 8= 7 + 3 + 1.

Variations

Line 5 can be used with the following variations:

> ➢ **Snatching.** When younger players are playing with grown-ups, the grown-ups can play with cards facing up and laid own on the table. Younger players can in their turn snatch a card from a grown-up player thereby having a larger choice of cards to create equations from.
> When playing with Snatching, all players still start a turn with at least 5 cards.
> ➢ **Advanced operators**, see the section Variations under Duel.

*The player lays down two number rows 3 * 3 = 9 i.e. 3 3 9 and 9 = 3 + 6 i.e. 9 3 6.*

The game finishes when both players cannot lay down more number rows. The winner is the player with most cards.

Card Game: 45er

Number of players: 2 players.

Preparing to play

> ➤ Firstly, time for a little creativity! To prepare for this game take a blank piece of A4 paper, turn it horizontally and write UP in the top middle and DOWN in the bottom middle as shown on the photo below. This is your **45er board**.

Create your own 45er board by writing Up and Down as shown above on a piece of A4 paper.

What to do

> ➤ In this game the first player lays down a number row of 4 hidden numbers, which represents a mathematical equation. The second player has to guess the equation and the 4 hidden numbers by laying down number rows.

Maquator 90

Which operators to use

The following operators and symbols can be used when forming equations using the cards 0 to 9:

- ➤ + i.e. addition.
- ➤ - i.e. minus.
- ➤ * i.e. multiply.
- ➤ / i.e. divide.
- ➤ = i.e. equal sign.
- ➤ ( ) i.e. brackets - optional.

How to Play

- ➤ The first player takes the 45er board and on the horizontal side (UP being in the top and DOWN in the bottom) lays down 4 cards facing down representing a mathematical equation. The first card is placed in place number one to the far left, the second card in place number two, third card in place number 3 and fourth card in place number 4 to the far right.

The first player lays down 4 cards representing an equation. The player then passes on the board to the second player.

- ➢ The remaining cards are left in a stack for the second player.
- ➢ The first player now turns the 45er board to the second player so the player has the 45er board in front with UP being in the top and DOWN in the bottom.
- ➢ The second player now lays down a number row representing a mathematical equation below the 4 cards placed on the 45er board. The **second player also calls out loud which mathematical equation the number row represents**.

The second player has the first try and lays down 15 = 9+ 6
i.e. 1 5 9 6.

- ➢ The first player now gives feedback to the second player's number row by:
  - o Moving any cards in the row under the 45er board which are **not represented** in the hidden row on the 45er board to the right of the 45er board.
  - o Leaving a **card** face up in the row under the 45er board which is **represented** among the 4 hidden cards on the 45er board (**either in the right or wrong location**).

○ **Turning any card** face up on the 45er board which is in the **right location** when compared to the number row laid down below the 45er board.

➢ Once feedback is given the game continues by the second player laying down another number row under the 45er board.

➢ The first player now gives feedback on the second row following the same guidelines as above i.e. moving any cards to the right of the 45er board if they are not represented on the 45er board, leaving all cards which are represented on the 45er board and turning the (unturned) card(s) on the 45er board which is represented in the second player's number row and is in the right location.

➢ The second player has 5 attempts to find the numbers on the 45er board and guess which mathematical equation the number row represents.

The first player provides feedback on the second player's first try. Above 9 is represented in the hidden 4 cards but the card is not in the right location since the equivalent card on the board is not turned. The 1, 5 and 6 are placed to the right as they are not represented in the hidden cards.

In the second try the second player moves the 9 then lays down the equation 4 * 9 = 36.

The first player gives feedback – the 9 is in right place and the equivalent card on the board is turned. 3 is kept, 4 and 6 placed to the right.

The second player now tries another equation in this case 3 + 9 = 12 i.e.
3 9 1 2

Now 3 and 9 are in the right place, 2 isn't and 1 is not used.

The second player now tries 3 = 9 − 2 − 4 i.e. 3 9 2 4.

3 cards are now in the right location but one card remains hidden.

The second player rearranges the cards to 3 * 9 = 27 and shouts out the equation.

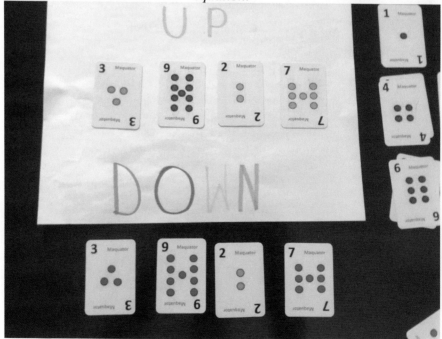

All four card are in the right location and the equation has been found.

Maquator 90

How to win and lose

- ➤ The second player wins by guessing the numbers on the 45er board, their right location and the mathematical equation the numbers represent in 5 or less attempts.
- ➤ If the second player guesses all the numbers on the 45er and the right location, but not a mathematical equation, which represent the numbers it is a draw.
- ➤ In all other cases the first player wins.

Variations

- ➤ **Hidden numbers.** The game can be played with 3 hidden numbers on the 45er board for younger players or 5 hidden numbers on the 45er board for more advanced players.
- ➤ **Advanced operators**, can be applied when playing, see the section Variations under Duel.
- ➤ **45er with colours.** 45er can also be played using the coloured dots on the cards. If played this way, the first player lays down a colour combination on the Maquator board. It is now up to the second player to guess the colour combination. The same rules apply as in 45er with numbers i.e. that a correct colour in the right location means that a card on the Maquator board will be turned. A coloured card which is not on the Maquator board will be moved to the right, while a card which is represented among the cards on the Maquator board but not in the right location will be left under an unturned card on the board.

Card Game: Equate Solitaire

Number of players

> ➤ 1 player

Preparing to play

> ➤ Turn one card face up on the table. This card is known as the **original face up card**.
> ➤ Turn a further 5 cards face up as shown below. These cards are known as **pool cards**.

The original face up card is placed face up with a row of 5 pool cards.

What to do

> ➤ Using the 5 pool cards, connect number rows representing mathematical equations to the original face card, or subsequently to cards connected to the original face up card.

Maquator 90

Which operators to use

The following operators and symbols can be used when forming equations:

- ➢ + i.e. addition.
- ➢ - i.e. minus.
- ➢ * i.e. multiply.
- ➢ / i.e. divide.
- ➢ = i.e. equal sign.
- ➢ ( ) i.e. brackets - optional.

How to lay out a Solitaire

- ➢ The player lays down a number row representing an equation using the original face up card.

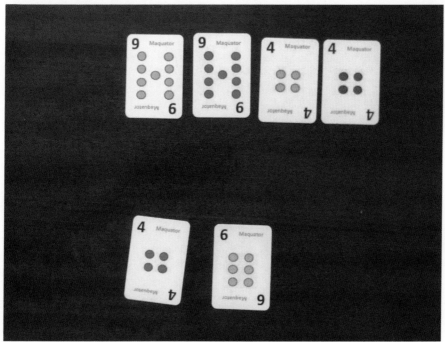

9 / 9 = 4 / 4 is laid down i.e. 9 9 4 4.

- ➢ After cards have been placed as part of a number row, turn cards from the stack to start a new turn with 5 cards.

> ➤ Use the pool cards and any of the connected cards already laid
> down to create a new number row representing an equation.

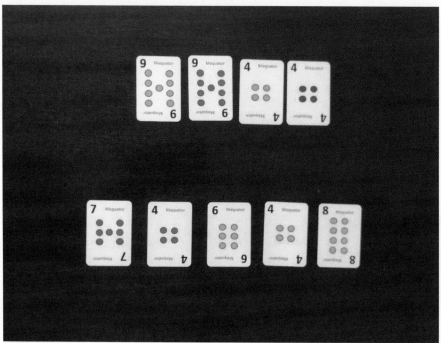

Each turn starts with 5 pool cards.

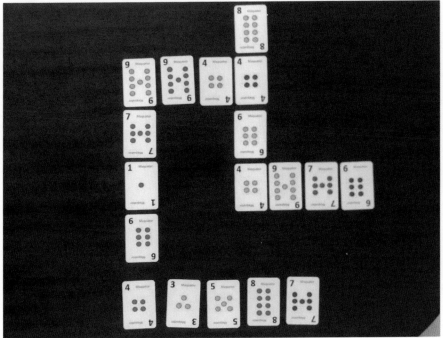

After a few rounds of play the player has connected other equations e.g.
9 + 7 = 16 represented by 9 7 1 6 and 4 + 9 = 7 + 6 represented by
4 9 7 6.

> If a number row cannot be added, turn another card to start the turn with 6 pool cards.
> The solitaire finishes when no further number rows can be laid down.

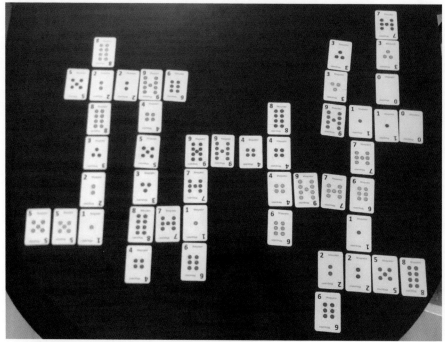

The Equate Solitaire completes when all possible cards have been connected representing equations.

Variations

➢ When 2 players lay out a solitaire then each player can take turns to lay out a number row from the 5 cards laying face up at the start of each turn.

➢ **Advanced operators** can be applied, see the section Variations under Duel.

FAQ and ..*a couple of "Don'ts"*

- ➤ The result of the equation can be written either on the left or right side of the equation's calculation.
- ➤ Numbers representing an equation are written either horizontally (left to right or right to left) or vertically (top to bottom or bottom to top).
- ➤ You need minimum one set of playing cards, but you can choose to play with more than one set of cards to prolong each game.
- ➤ In Duel, Line 5, 45er and Equate Solitaire:
  - ○ A single standing 0 is not permitted e.g. $1 + 0 + 1 = 2$ represented by a 1, 0, 1 and 2. Similarly multiplication with 0 e.g. $0 * 8$ is not permitted.
  - ○ A player cannot extend/add to an existing number row.

Appendix:

You can create a set of Maquator 90 cards for example using a card board box.
<u>Warning - if you are not familiar with using scissors do not try this!</u>

1. You can take a cereal box, gentle open it where it is glued together and from the card board cut 50 equally sized rectangles. We recommend a size of 4 cm wide and 6 cm high.
2. Now create the Maquator cards by writing 0 to 9 in the corners of the cards as shown on the photo below. You should have 5 0 cards, 5 1 cards, 5 2 cards and so on up to 9.
3. Then draw dots on each of the cards 1 to 9 representing the numbers on the cards i.e. draw one dot on cards with 1, two dots on cards with 2 and so on up to 9.
4. Finally, take all the 1 cards and colour the dots on the card in a different colours. Then take all the 2 cards and colour the dots each in different colours and so on up to 9.
5. You should now have 50 cards with 5 0 to 9 cards with in total 9 yellow, 9 red, 9 blue, 9 green and 9 pink and 5 0 cards (blank).

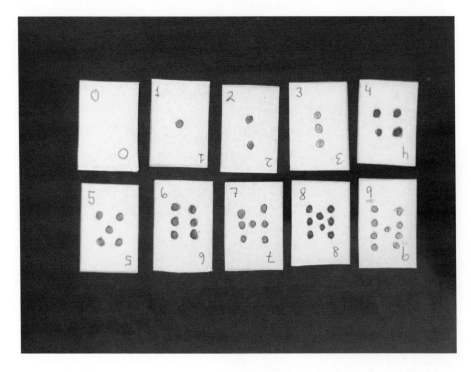

Printed in Great Britain
by Amazon

33403485R00025